The Rooster's Wife

T0131227

BOA
EDITIONS
LIMITED

The Rooster's Wife

Poems by Russell Edson

AMERICAN POETS CONTINUUM SERIES, NO. 90

BOA EDITIONS, LTD.   ROCHESTER, NY   2005

First Edition
05 06 07 08   7 6 5 4 3 2 1

Publications by BOA Editions, Ltd. — a not-for-profit corporation under
section 501 © (3) of the United States Internal Revenue Code — are made
possible with the assistance of grants from the Literature Program of the
New York State Council on the Arts; the Literature Program of the National
Endowment for the Arts; the Sonia Raiziss Giop Charitable Foundation; the
Lannan Foundation; the Mary S. Mulligan Charitable Trust; the County of
Monroe, NY; the Rochester Area Community Foundation; the Elizabeth F.
Cheney Foundation; the Ames-Amzalak Memorial Trust in memory of Henry
Ames, Semon Amzalak and Dan Amzalak; the Chadwick-Loher Foundation in
honor of Charles Simic and Ray Gonzalez; the Steeple-Jack Fund; and the
CIRE Foundation, as well as contributions from many individuals nationwide.

See Colophon for special individual acknowledgments.

Cover Design: Steve Smock
Cover Art: "Woodcut" by Russell Edson
Interior Design and Composition: Scott McCarney
Manufacuring: McNaughton & Gunn
BOA Logo: Mirko

Library of Congress Cataloging-in-Publication Data

Edson, Russell.
   The rooster's wife : poems / by Russell Edson.— 1st ed.
   p. cm. — (American poets continuum series ; v. 90)
   ISBN 1-929918-62-3 (alk. paper) — ISBN 978-1-929918-63-8
   (pbk. : alk. paper)
   1. Prose poems, American. I. Title. II. Series.

   PS3509.D583R66 2005
   811'.52—dc22

2004024831

BOA Editions, Ltd.
Thom Ward, Editor
David Oliveiri, Chair
A. Poulin, Jr., President & Founder (1938–1996)
260 East Avenue, Rochester, NY 14604
www.boaeditions.org

NATIONAL
ENDOWMENT
FOR THE ARTS

State of the Arts

NYSCA

for Frances

CONTENTS

≫ 1 ≪

≥ 3 ≤

Fairytale

Behind every chicken is the story of a broken egg. And behind every broken egg is the story of a matron chicken. And behind every matron is another broken egg. . . .

Out of the distance into the foreground they come, Hansels and Gretels dropping egg shells as they come. . . .

History

Structure and sense that dreams from the corners of a room. . . .

Table edges that remind us of tensions drawn from exacting boundaries, falling finally up from the patterns of a rug. . . .

Those drapes hanging by that window, draped like classical stone, shifting with subtle compliance to an atmosphere softly breathing from a distant meadow. . . .

Above us a monstrous artifact of clouds that've lain together for centuries like sleeping swine. . . .

The drift. . . .

The Wonders of Nature

A circus manager, who secretly likes to wear women's clothes, has run out of money and is selling his wonders-of-nature show.

A slightly damaged fat lady, who for lack of a watercress salad has lost a couple of ounces of carefully nourished heft, priced for quick sale.

A contortionist who has twisted himself into an emotional knot being offered as a piece of modern sculpture.

A special bargain, Siamese twins, buy one and get one free. Two for the price of one.

A three-legged man who has only two, but insists on a third; you have only to open his fly—By appointment only. Ladies preferred.

Finally, the bearded lady, who is actually a man wearing a dress. Otherwise, the circus manager himself with a goatee pasted on his chin. . . .

The Mare's Egg

The mare had layed an egg. It must have been the rooster made love to her.

Would you call what animals do making love?

He probably flew up to her chestnut rump and perched there with his bird feet, arching his posterior to the task.

But would you really call that making love?

Does it matter?

Of course, dear reader, whether animals can be said to make love or not, when the mare's egg opened a small pink Pegasus flew up and perched on the chestnut rump of its mother. . . .

Evenings

There was a man who danced with his dog; his wife loving to sit evenings watching them. . . .

Suddenly she screams, Stop!

What happened? cries her husband skidding to a stop; the dog fainting.

I don't know, she says, Suddenly everything's reversed, now I don't love sitting evenings watching them.

Who?

That man who dances with a dog; didn't you see them? They were here just moments ago.

Then the man begins dancing with his dog again. . . .

Suddenly his wife screams, It's reversed, now I love sitting evenings watching them.

Who? cries her husband skidding to a stop; the dog fainting.

That man who dances with a dog; didn't you see them? They were here just moments ago. . . .

The Dog's Tail

An old woman was absentmindedly stirring a pot with a dog's tail.

When her husband asked her about the furry stirrer she said, It's the dog's tail, it came off in my hand.

When her husband asked her what she was stirring she said she didn't know, that all her thoughts were now for the dog's tail.

When her husband noticed that the dog was in the pot she said, Oh, is that where he is? I wondered where he got without his tail.

Her husband said, I'll bet he likes that, being stirred with his own tail. It's sort of like the tail wagging the dog.

The old woman said, I was just petting him, and it came off in my hand. I hope God wasn't looking.

Her husband suggested that perhaps it wasn't the dog's tail that broke off, but rather the dog that broke off. . . .

The Elegant Simplification

An old man's cane had broken a bone. Actually, a cane has only one bone. One of nature's more elegant simplifications.

As the doctor prepared the splint he asked how the cane came to break its back.

My wife, said the old man, Her head is uncommonly hard. . . .

The Tree

They have grafted pieces of an ape with pieces of a dog.
Then, what they have, wants to live in a tree.

No, what they have wants to lift its leg and piss on the
tree. . . .

The Proud Citizen

An old man was proud that he had passed his years, as he had his breath and stools, without his having killed anyone.

He wondered if he might not report this to the police, hoping to be received with sirens and blinking lights of penal gratitude.

He would explain that he had had many opportunities, that it wasn't just laziness; that virtue without lure of sin hardly earns its name. . . .

Super Monkey

He was creating a super monkey by grafting pieces of a dead parrot to a morphined monkey.

When the monkey awoke he was covered with green feathers and had a beak. His first words were, Polly wants a cracker.

It's historic! No monkey will ever have said this before!

And so super monkey will be given all the crackers super monkey can eat, until super monkey sickens of crackers and says, Polly wants a banana. Which will be another historic quotable!

Then he'll begin work on superduper monkey who, with proper grafting, will be able to sing like a canary. . . .

The Bloody Rug

A daughter was beating up her father. . . .

Have mercy, child, cried the bloodied father, I might, in defending myself, have unintentional contact with your parts; my fingers tangled unintentionally in your underclothes.

Don't you touch my parts with all your fingers tangled in my underclothes, she screamed.

The mother looking up from staring at the rug said, Why are you beating dad?

To keep him from tangling his fingers in my underclothes.

That's a lie, cried the father, She suddenly attacked me, and in defending myself I feared I might have unintentional contact with her parts, which is against the law; my fingers tangled unintentionally in her underclothes, which is also against the law.

The mother looking up again from staring at the rug said, Please don't beat dad.

Why? said the daughter.

Because you're getting blood on the rug. . . .

The Courtship

A woman wanted to sell one of her knuckles. . . .

Tell me, she said, How many carats do you think it is?

But it's not a jewel, said the jeweler.

Nor are you, said the woman, Much less a gentleman.

If you insist, said the jeweler as he put her knuckle to his loupe. Hmmm, he said, It's certainly not a clear stone, a lot of cracks. I hope you didn't pay too much for it.

It was a gift, mother gave it to me. What do you think it's worth?

Not very much, said the jeweler.

Maybe the setting has some value? said the woman.

It might, said the jeweler, If it includes two tits and a cunt. . . .

I love you, said the woman.

Let's get married, said the jeweler.

You go too far, said the woman. . . .

The Guardsman's Horse

A hungry king said he was so hungry he could eat a horse. So a guardsman killed his horse and fed it to the king, who said, Did you feed me a horse?

Yes, Sire, you said you were so hungry you could eat a horse.

But only figuratively; now I'm full of horse. Did I eat the hooves?

Yes, Sire, the mane and tail, too.

Did it taste good?

I don't know, Sire, I fed it all to you.

But it was your horse. If a guardsman doesn't know what his horse tastes like, who does?

The Bleeding

Young Harry was an excellent bleeder. As he got older he became even better, his skills expanding to even greater flood.

They said, Harry, you're really ready.

For what? said Harry.

To launch your career.

So Harry began to bloody himself all over the world. Bleeding in the finest theaters, and in command performances for the crowned heads of Europe, many of whom still retained fond hemophiliac memories.

Finally, after years of bleeding, his blood grown thin and translucent, they began to boo him. His career done, he retired to a little cottage to live out his remaining days in quiet anemic anonymity.

Oh, yes, he would sigh, There was a time when my blood was as thick and as red as ketchup, and just as tasty. . . .

The Running of the Mascara

A woman weeps as she reads an old love letter written by her grandfather to his automobile. Her mascara runs as though she had ink instead of tears.

Why are you crying, Mildred? asks her grandfather.

This beautiful letter, she sobs.

Then her grandfather reads it and begins to cry. First with heavy, manly sobs choked in his breast. Then, fully melting he cries with the unguarded openness of a child. Finally he begins to scream in a high-pitched falsetto.

But why are you crying, grandfather?

This beautiful letter, he sobs.

Then they read the letter again and once more begin to weep. He in high-pitched falsetto, she in woman's scream. Her mascara branching out of her eyes like a black lightning streaking the room.

Dearest Automobile, I love you better than my parents. Mom's been nice, and dad gave me a fishing rod. But, hell, it's you I love. If you feel the same, I'll sneak out tonight, and we can make love in your back seat. . . .

Things

A woman said that she had been feeling rather thingy of late, and had, in fact, just had a thing.

Her husband said, What thing?

That thing, it came out of my body as suddenly as not, like some presumptuous stool.

That's a baby, you silly thing.

Then is it to suckle at my thing?

Where else? you silly thing.

So the woman held the baby to her chest and said, Now the thing is attached to my thing. But I don't know if the thing is suckling my thing, or my thing is suckling the thing.

Her husband said, It's the thing suckling your thing, not your thing suckling the thing, you silly, silly thing. . . .

The Cat

An old man is dressed in lingerie to amuse his cat. He says, You didn't know your papa was a lady, did you, pussycat?

Then being visited by another old man he says to the other old man, You're probably wondering why I'm wearing lingerie.

Not at all, says the other old man.

Surely you must be curious, perhaps even a little disgusted at seeing an old man romping about in silk undies like a sex kitten.

Well, now that you mention it. . . .

— Yes, let's put everything on the table.

In that case, why in hell are you wearing women's underwear?

To amuse my cat.

To amuse your cat. Why didn't you say so, that explains everything. For a moment I thought you had lost your mind. Incidentally, where is your cat?

I don't have a cat. . . .

The Hollow Pig

A butcher had hollowed out a pig to make himself a pig costume. Then he had a pile of pig insides.

As he crawled into the hollow pig he was trying to have pig thoughts to complete his costume, and at the same time trying to think of what to do with all the pig's insides.

He thought he might hollow out another pig and stuff it with the first pig's insides. But then what to do with the insides of the second pig, hollow out a third?

No, no, for once started he'd have to hollow out every pig in the world, and still wind up without a place to put the insides of the last pig. . . .

The Great Abstraction,
or the Unfinished Story of Everything

A woman had a bowel movement. An attending doctor smiled and said, Yours will be the first baby born without a belly button.

Does that mean I've won a prize? said the woman.

I'm not sure, said the doctor.

Not sure? Then why did I bother? said the woman.

Because in the final accounting, dear lady, things happen in abstract rendezvous as fate and substance meet to tell the story of everything.

Does that mean I don't get a prize? said the woman.

I'm not sure.

When will you be sure?

When fate and substance meet, and the full story of everything has been written.

When will that be?

When fate and substance meet, and the full story of everything has been written.

And when will that be?

Hopefully any day now. . . .

The Way

Now that you are falling down your stairway, what is it you have forgotten? Is it a window where last you saw the sky, that area above Monday morning's earth?

Now that each stair is whacking you back, breaking your calcium tree, you would have thought then to have walked more carefully. . . .

The Jack Story

There was the Jack of the beanstalk story, and a Jack Sprat who could eat no fat. And there was Jack-in-the-box who used to spring out of a box for no reason at all. And Jack who broke his crown fetching water with a certain Jill. Not to forget little Jack Horner, or the Jack who jumped over a candlestick. . . .

Theirs is a club of Jacks. Grown old they are all drunks. Jack Sprat's a bloated sot. Jack of the beanstalk has long ago drunk up all his beanstalk wealth. Jack who used to spring out of a box now lies at the bottom of it in his own vomit. Little Jack Horner just sits in a corner nursing a bottle of rye, saying, What a good boy am I. And the Jack who used to fetch water complains that he still misses Jill, and all the wondrous falling they used to do. . . .

Baby Doll

A dying old man is presented with a toy coffin made of cardboard.

A cardboard coffin? sighs the old man.

It's a funeral toy for little girls who like to play pretend death with their baby dolls.

But why cardboard? he sighs.

It's cheaper than wood, and good enough for make-believe funerals.

But what will my friends think when they see me laid out like a baby doll in a coffin made of cardboard?

You have no friends.

Then what will the God think when I arrive in a coffin made of cardboard?

There is no God. . . .

The Womb

There was a man who built himself something that looked like a womb. When he left it he'd say, I've just been born, spank me if you must, but please be kind.

And just before he entered it he would say, I'm just about to be conceived. I wonder who my daddy and mommy are?

To Do No Harm

A doctor, keeping to his promise to do no harm, keeps a spare old man with a white beard in his medical closet in case someone should choke to death on a tongue depressor. Then he presents the spare old man to the patient's wife, in lieu of her husband, as another miracle of modern science—No extra charge for the beard.

Or, say he accidentally cooks a little girl to death with his x-ray machine, he can hide her under his white jacket and present the spare old man to the mother waiting in the waiting room, explaining that as he cooked her little girl she suddenly went into puberty, sprouting all kinds of secondary hair and nipples—Look, she even has a beard like a billy goat.

But even so, say the doctor should accidentally cut his own throat while shaving with his scalpel (this rarely happens), then he collapses into a pool of his own blood wondering if there is anything past death. If not, he simply fades into what he was before his mom and dad had groped each other in the dark.

And still no harm was done. . . .

Vignette

A man inside one distance looked down into another and saw his mother, and climbed down through the bric-a-brac.

But when he got to her she was still as small as she was in the distance.

But why are you still so small? he cried.

It's you, she cried, You're still in your own glutton foreground, optical spendthrift!

But mother aren't you supposed to swell to size when I come to you?

I don't swell anymore, I'm too old. You'll have to find somebody else, she cried as she got even smaller. . . .

At Sea

Two wifeless men at sea fishing for mermaids to wed. . . .

They wonder if their children will be fish. If daughters, they'd like them to be Emily Dickinsons born with water wings; nice shut-in types who write poetry and love their dads.

They're not against incest as long as it's kept in the family. Didn't Adam even make love to a piece of himself and create a whole species?

At last one of them says, Here we are at sea, ready to be fathers, and not one mermaid to accept our milt.

Perhaps we should just make love, as did Adam, to our ribs—The floating ones, which seems more than right for men who find themselves at sea. . . .

The Civilized Man

When going out he would wear handcuffs in case he committed a crime; a good citizen ready to be arrested. Okay, officer, you've got me dead to rights. Perhaps you'll want to give me a good whack on my head with your nightstick? Well, go ahead, until proven innocent I shall remain guilty, leaving it in your hands to prove my innocence.

Perhaps I shall be proven criminally insane. Then I shall demand a straitjacket as my democratic right, and take up residence in a room made of rubber.

Sometimes this is the only thing left to a civilized man. . . .

The Wet Diaper

Rocking the baby in her arms its head falls off and rolls to the corner of the room and begins to cry. Of course being separated from its lungs it cries without sound. So the mother, still attached to her lungs, begins to cry.

The father, hearing her, comes running, asking, What's wrong with the baby?

It wet its diaper.

Oh, good, says the father, I thought its head might've come off. It looked a little loose this morning. . . .

Breakfast Toast

As a man watched, his wife buttered his hand.
He asked her why she was buttering his hand.
She said, I thought it was a piece of toast.

When she bit his hand he asked, Why are you biting my hand?
I thought it was a piece of toast.

When she bit his hand again he asked, But why are you still biting my hand?
Because I still think it's a piece of toast. . . .

The Grandpa Knee

An old man who was old enough to be his own grandfather said to himself, Grandpa, may I sit on your knee?

And replied, Sit on your own knee, you're old enough to be your own grandpa.

But, grandpa, it's the grandchild who sits on the grandpa's knee.

Grandchild? Why, you're old enough to be your own grandpa.

But, grandpa. . . .

—But grandpa nothing! There's no sense to it, one grandpa sitting on another grandpa—It's redundant!

But, grandpa. . . .

—If you don't stop grandpaing me I'll put you across my knee and give you a good spanking. . . .

The Knitting

An old woman was knitting herself a pair of gloves.

She said as she stared at a distant cloud no bigger than a hand floating in her window, Here I knit a pair of hollow hand-shapes with the very hands that will wear them. . . .

Please have mercy, sighed her husband.

Does that give me permission to knit a pair of socks with my feet? she said.

Perhaps a brassiere with your breasts, he sighed.

—A jockstrap with your pissed-out testicles, you cranky old man!

All I ask is a little mercy. . . .

Of Those Who Bring Forth
Upon the Earth

A woman had given birth to twin daughters, but used only one.

When her husband asked where the other one was, she said, There's only one.

But I thought I once saw two partially developed persons of the female persuasion exit the lower end of your torso.

That was when I was feeling excremental and lay down in expectation of an extraordinary bowel movement, but instead bore a partially developed person of the female persuasion.

But I thought I saw two partially developed persons of the female persuasion exit the lower end of your torso, said the husband.

Oh, that one, she's a spare in case I come again into expectation of an extraordinary bowel movement, but find my bowels unwilling; then would I tender the spare partially developed person of the female persuasion, and so remain within the index of those who bring forth upon the earth. . . .

New Prose About an Old Poem

One day an old poem is carried away by the wind. Its poet is relieved, now he won't have to be nice to it anymore.

The poem was always too good to throw away, yet, not good enough to publish.

It lived with him demanding to be reconsidered every so often.

But, even so, he sees that he's not to be rid of the old poem, the wind in reverse has returned it to his desk.

The old poem is glad to be home, and wants to be read again.

The poet reads it and realizes once again that it's too good to be thrown away. Perhaps, he thinks, he'll send it out in the next mail, knowing, of course, that he won't; and that he'll have to go on being nice to it for the rest of his life. . . .

Geography

A young scholar is hired to tutor the daughters of Mr. and Mrs. Father and Mother. But somehow his hands are always up under their dresses and down into their underpants.

Mr. Father says, I wish you wouldn't do that.

Do what, Mr. Father?

Put your hands in my daughters' underpants and touch their genitalia. . . .

But one day while teaching geography he had undressed one of Mr. Father's daughters and was cupping her mons veneris with one hand and one of her small breasts with his other hand.

When Mr. Father looked in he said, I thought I mentioned something about your not touching my daughters' genitals. Mrs. Mother feels the same way in regards to our daughters' genitals and secondary sexual characteristics.

Oh, that, said the tutor, I was just showing the girls how the science of geography began.

But what has my daughters' genitals and secondary sexual characteristics to do with geography? said Mr. Father.

Geography begins at anatomy, said the tutor, I was showing the girls how the urge to explore was first awakened by the female form. Ergo, the female figure being the model and inspiration for the first explorers; the reason why the planet is called Mother Earth. Then of course came the pornographers who, as it turns out, were the first cartographers.

Forgive me, I didn't mean to interrupt anything so intimate as geography, I should really wear a bell, said Mr. Father, But they're so noisy and, as you know, Mrs. Mother likes to sleep late. . . .

Meanwhile, the tutor had undressed Mr. Father's other daughter and was kissing her just under her left ear. . . .

Of a Midsummer's Night

As I was nearing earth out of the sky, returning from the moon, a basket of mushrooms on my arm, I saw a village that looked like a toy.

Mushrooms are grown in the craters of the moon by moon dwarfs who, of course, have moon-faces cratered with the scars of pubescent meteors; their eyes as dark as the dark side of the moon.

And as I rode a moonbeam down, a basket of mushrooms on my arm, what had seemed a toy village was indeed a toy. And there an old woman, as small as a mouse who looked like my mother, smiled at me as if she thought me to be the man-in-the-moon.

But as the dark of my descent engulfed her she suddenly scampered away in fear, her Mother Hubbard billowing, her tail following. . . .

The Sweetheart

An old woman had fallen in love with one of her feet.

Her husband said, No you didn't.

Yes I did, it was sticking out of the covers of my bed, and I said, You're a sweetheart.

No you didn't, said her husband.

Yes I did.

No you didn't.

It was sticking out of the covers of my bed, and I said, You're a sweetheart, so calloused for the many roads you walked to find me; when in truth, I was right here all the time, you had only to crawl up my leg to find me. . . .

Lunch

An old man lies in the mud with closed eyes, nursing the teat of a sleeping sow.

His wife crawling out of the pig shed on all fours says, Oscar dearest, are you dead?

Their son hobbling out of the shed, dragging his trousers from his ankles, asks, Has dad become a piglet?

Their daughter crawling out from under the shed wearing a Hitler mustache of pig manure asks, Has papa fallen in love, or is he having lunch?

Who knows, sighs the old woman, It could probably go either way.

But what if it's love?

We must be careful, sighed the old woman as they knelt in the mud for lunch; the sleeping sow mumbling nursery rhymes in pig Latin. . . .

The Shrieking

A man who had found something to love ran shrieking to his mother's bed.

What wonderful news, she shrieked.

I'm tired of sleeping alone, he shrieked.

Of course you are, because you're a normal type person. Who's the lucky girl? she shrieked.

My bed, he shrieked.

My fondest wish, she shrieked, It was always my hope that you'd take to your bed one day.

I hope you don't mind that we've already sort of slept together? he shrieked.

Of course not, why, she's almost like one of the family—Like you, she shrieked.

Then it's okay? he shrieked.

Yes, okay, anything—Just stop shrieking, she shrieked, You're hurting my ears. . . .

Of Baskets and Eggs

There was a woman who had been advised not to put all her eggs in one basket; that it might just be that a dozen eggs require a dozen baskets.

Her father said, Please, you're making me nervous.

Would you say that the advice is more figurative than practical?

Her father said, Please, you're making me even more nervous.

I'm undecided about the ratio of eggs to baskets in reference to the transference of same, say, from nest, A. to point, B.

From nest, A. to point, B.? Now I'm really nervous, said her father.

Nervous? What's that?

Impulses without expression. Regret and anger battling through the system. Hatred compressed under a father's patience, said her father.

But what's that got to do with the question of baskets and eggs?

The Fascination

A rat was trying to fit its tail into an old woman to keep it from being stepped on.

Don't, said the old woman, Not at my age.

The old woman's husband said, The rat's trying to do with you as I did when I was doing with you. It's fascinating.

It is not fascinating, cried the old woman, No more so than when you were having to do with me.

But, as the old husband and wife disputed the fascination of it, the rat fitted its tail into the old woman.

When they discovered the rat doing as the old man had, the old woman said, See, it is not fascinating.

And the old man said, You're right, it is too biological. . . .

Bath Water and Spilled Milk

A woman had thrown her baby out with its bath water. Her husband was in the kitchen crying over spilled milk.

The woman said, I don't know where the baby is, the last time I saw it I was throwing out its bath water.

I know, but this is even worse, wept her husband, All this milk on the floor.

The last time I saw it, said his wife, It was in its little tub. I had just finished bathing it and was throwing its bath water over the rail of the back porch. . . .

Yes, of course, but how will I ever get all this milk back into its bottle? wept her husband.

Well, it does no good crying over spilled milk, said his wife.

Why not? It's just as good a cliché as yours, screamed her husband. . . .

The Box

A man was opening and closing a box. Once it was opened he closed it. But again, closed, he would open it—Opened, he would close it—Closed, he would open it. . . .

And he continued to open and close the box. Opening it and then closing it. Opening it and then closing it. . . .

The constant motion made his mother nervous. She said, Will you stop doing that with that box?

Okay, I'll just close it, he said.

But no sooner was it closed than he was opening it again.

I said stop doing that with that box, she cried.

Okay, okay, I'll just open it and leave it open, he said.

But no sooner was it opened than he was closing it again.

His father said, Why don't you get married so you don't have to be doing that with that box?

But I am married, this is my wife, said the man.

Congratulations, said his father, But why didn't you tell us you were married?

Because I thought you might be a little disappointed that I didn't marry the girl next-door, said the man.

But there's no girl living next-door, said his father.

I know, that's why I didn't marry her. I hope you're not disappointed.

Not at all, said his father, It's just that that box is a little disappointing. . . .

The Wound

A womanness had formed in a man's hand, which he called the wound of his desire. . . .

He asked his father if it was a good thing that a man marry his hand.

Marry your hand? cried his father, Then what will things have come to when men have married their hands?

The intimacy already speaks to the conjugal, said the man as he showed his father how it was with his hand.

Your hand is full of womanness, cried his father, It is not right that men look upon unclothed womanness without that she were clothed in man's desire, lest it were that woman was to judgment come; nor right that men take their hands in weddedness, that they become one-handed to the true work of the world. . . .

Baby

After nine months a woman gives birth to a little girl's doll.

The doctor's a little disappointed about the baby not having an umbilical cord, which he takes special delight in cutting on newborns.

The mother and father decide to love it and take it home. They put a diaper on it and give it an empty baby bottle to nurse. And life is good.

This is what it's all about, says the husband.

What what's all about? says the wife.

You and I, and the fruit of our loins. . . .

But one day the baby begins to make a ticking sound. They call the bomb squad, and the baby is put in an explosives cage and taken to a deserted field, and blown-up. . . .

Let Us Consider

Let us consider the farmer who makes his straw hat his sweetheart; or the old woman who makes a floor lamp her son; or the young woman who has set herself the task of scraping her shadow off a wall. . . .

Let us consider the old woman who wore smoked cows' tongues for shoes and walked a meadow gathering cow chips in her apron; or a mirror grown dark with age that was given to a blind man who spent his nights looking into it, which saddened his mother, that her son should be so lost in vanity. . . .

Let us consider the man who fried roses for his dinner, whose kitchen smelled like a burning rose garden; or the man who disguised himself as a moth and ate his overcoat, and for dessert served himself a chilled fedora. . . .

The Organ of Thought

A man thought long of the organ where he thought long of it. A pupa, he thought, his skull its chrysalis.

At death it would find its wings in imago, he thought. Then he was free and his angel let go.

But, oh, he was so tired of thinking. . . .

Night Watch

One night a man in a dark suit almost as dark as the night, turns to another in a suit almost as dark as the night, and says nothing.

The other turns and answers with silence. . . .

But the hour grows late, time to brew coffee, and time to light cigarettes, even as the windows gray with the first fog-moist kisses of the dawn. . . .

The Dark Waters

Behold the clam, the hard lipped mouth in perpetual smile, as if embarrassed at the irony of its choice.

And mussels sleeping on the coast like faces that have disappeared into dreams, leaving only purple eyelids to guard their sleep.

And here jellyfish rise and fall in the tides like brains that have lost their minds.

And nowhere does the naked Venus rise. . . .

The Truffle Garden

A customs pig is sniffing the ears of a man who would sleep. . . .

Don't worry, Mr. Sleeper, he just wants to make sure you're not hiding anything in your ears like contraband truffles.

I can assure you, I keep nothing in my ears, except the inside of my head.

You don't mean to say there's an inside to your head?

Uh oh, the pig doesn't think you're all that serious.

About what?

The trip.

To where?

Dreamland, where else?

But I've made no plans to go abroad.

Oh, that makes a whole new thing of it. Now the pig thinks you are serious.

About what?

About entering the land of dreams.

Then the pig has found nothing that needs declaring?

Nothing perhaps, except that truffle garden buried in your head. . . .

The Moonlit Wall

Someone standing against a moonlit wall trying not to be seen. It won't work, he thinks, someone will see me anyway. . . .

And there you are, Mr. Shadow, standing against that same moonlit wall, still trying not to be seen.

Ah, yes, but you saw me even as I tried not to be seen. . . .

And there you are again.

No hope for it, you will see me no matter how many times I try not to be seen. . . .

And there you are again. Not very good, are you, Mr. Shadow?

No, you will see me against this moonlit wall no matter how many times I try not to be seen. . . .

The Gas Heads

A balloon of gas issues from a head named Jane, with words printed on it: Look, Dick, look at Spot, Dick.

And then another balloon of gas seeps out of the Jane-head: Look, Spot, look at Dick, Spot.

Dick looks at Spot, and Spot looks at Dick.

Now a balloon comes out of the Dick-head with printed material on it: Look, Jane, why must everybody look at everybody, Jane?

I don't know, Dick, it's just the way it's printed on our gas.

More gas: Look, Jane, look at me, Jane.

More gas: I am, Dick. So is Spot, Dick. Are you looking at me?—Here's looking at you. . . .

Skeleton

A skeleton awakens and yawns in a grinning gape of teeth; stretching, its spine arching like a suspension bridge.

Sitting on a coffin it pulls flesh as if hip boots up the bones of its legs. Toe bones wriggling and twisting to fit themselves into the flesh of their feet. And then a long sleeved glove of flesh up one arm, and then the other.

And so with its torso, wrapping its ribs in a corset of flesh, fastened with the stub of an old umbilical cord.

Almost complete it hangs its genitals and puts on a mask.

In the street it passes unnoticed by other skeletons also wearing masks. . . .

Rocks

Two old men were performing autopsies on each other. And as they worked, putting this in a glass jar and that in a chamber pot, they talked of rocks; arthritic rocks, and rocks with rotten teeth; rocks with gout, and rocks with bad stomachs; rocks with hair in their ears, and rocks with scrotums hanging to their knees; rocks with gall stones, and rocks blind with cataracts. . . .

Suddenly one of the old men says to the other, Did you know you were pregnant?

No, says the other old man.

Then holding up a rock, he says, Look what I found in your womb.

Spank it, says the other old man, And see if it cries. . . .

What It Is of the Wood

There was an old woman who would open her shawl and glide in a wood, squatting now and again on the branch of a tree to do urine.

Her husband asked her what it was of the wood.

I go to the bathroom there, the facilities are really quite modern.

But we have indoor plumbing, he remarked.

Indoor plumbing? Where?

In the bathroom.

In the bathroom? she squealed.

It's been there for years.

For years? Oh goodness, now I really have to pee, she squealed as she jumped through a window flying for the wood, trailing a long ribbon of toilet paper into the trees. . . .

The Occasion

A large female presence is floated under a helium balloon to a sofa.

A man without legs walks into the room led by a blind seeing-eye dog.

Another, without a mouth, begins to sing a duet with someone shouting, Fire! on a bullhorn.

Still another, with his penis dangling like a caterpillar from his fly says, I'm sorry to interrupt this august occasion, but has anyone seen my pet condom? It crawled off my weewee and might be trying for a metamorphosis, even though butterflyhood has been promised to my weewee.

Meanwhile the helium balloon begins to tug at the large female presence like a sleepy child wanting to go home. . . .

The Restaurant

A series of bangs, like backfires or gunshots. . . .

It was a fat man exploding on the floor of the restaurant, writhing and jerking with bright flashes tearing through his clothes.

Obviously too many explosives. One spark arcing an imperfect nervous system, and a man can be destroyed by his own bowels.

Soon, after a few smoky belches and our encouraging nods, he was studying the menu again, and ordering the gunpowdered chicken with the nitroglycerin sauce.

After it was over, as well it had to be, and the waiters had finished sweeping him up, I picked a piece of intestine out of my glass and resumed my Molotov cocktail. . . .

Thighs

There was a woman who did not admit to having thighs.

No thighs?

Not even a crotch, she smiled with pride.

No crotch?

Not the slightest hint of an anus or a vagina, she smiled again.

What about a slight rump crack?

Not even a belly button, nor the slightest hint of a Shirley Temple dimple, she said, smiling even more broadly.

By this time everyone had fallen madly in love with her, and with pieties stiffened longed to kneel between her thighs. . . .

Perhaps it was her smile?

Dessert

Dr. Lully was enjoying a crushed fruit dessert that had come to his table in an ambulance.

A cleaning woman, seeing the Doctor's labor, began to wipe his brow with her underpants, which she had just slipped off to service that same brow.

Don't you have anything else to wipe my brow with? asked the Doctor.

I was just trying to help.

I've never heard of anything so foolish as a nurse using her underpants to wipe a doctor's brow, said the Doctor.

I was going to use my wet mop, but it just seemed too wet. I mean your brow didn't seem as wet as the mop, and I thought it might make your brow even wetter than it was.

That makes sense, said the Doctor as he finished his dessert.

May I put my underpants back on?

By all means, unless you're prepared to have a baby. Who knows what interesting things have come into your bared lower parts? But don't look at me, I've already had my dessert. . . .

Looking for the Head

A woman had given birth to a small pink elephant.

She asked the doctor, Why an elephant and not a parrot? Isn't your husband an elephant?

No, that's grandpa Tusk. My husband's the parrot, the one in the cage. You met him when you came to look under my skirt. I said, Why are you looking under my skirt? You said you were trying to see if you could see the baby's head. Meanwhile my husband was emptying his bowels in a newspaper on the floor of his cage. Don't you remember?

No, my head was under your skirt looking for a head. . . .

An Old Woman Who Tastes Like
an Old Woman

An old woman, afraid she might bite herself, hides her false teeth in a cookie jar before she sleeps.

And a nice nibble for the little dears. One of which she's sure is a sweet tooth.

But her mouth said, Don't be so precious, I wouldn't bite you for all the cookies in the world, you don't taste good.

I taste every bit as good as you, said the old woman to her mouth.

No you don't, said her mouth, You taste like an old woman. . . .

Portrait of a Wrinkly Old Man
with Nasty Genitals

There's this old grandpa dressed in a monkey suit, who wants to be known as grandpa monkey.

But you're a wrinkly old man with nasty genitals, says one of his descendants.

I have evolved, says the grandpa, Besides, I've fallen in love with a monkey maiden.

But you're too old to be in love.

I am not. I plan to live in a jungle with my monkey bride, and build a nest high in the trees and have love with her.

But you're too old to have love.

I am not. We plan to have love until the trees are filled with monkey children.

But you're just a wrinkly old man with nasty genitals wearing a monkey suit.

I am not. I'm a monkey who dressed himself in a monkey suit.

And with that he took off his monkey suit, and showed them that there was a monkey under his monkey suit.

Congratulations, you wrinkly old man with nasty genitals, you're a monkey!

Little did they know that the monkey under the first monkey suit was yet another monkey suit, and that under that was the same wrinkly old man with nasty genitals. . . .

Little Edward

As a man was playing through a particularly emotional passage his violin exploded like a breast broken by its own heart.

Why, you dirty little traitor, said the man, his hands full of kindling and gut.

His wife said, What did you do to little Edward?

Why are you asking what I did to little Edward, and not asking what little Edward did to me? cried the man.

Well, what did little Edward do to you?

He ruined my violin. . . .

Flies

There is nothing so disarming as a housefly, a six legged animal somewhat smaller than a horse, but a lot more massive than an amoeba. They prefer Musca domestica on formal occasions. But for every day use, housefly will do, it's friendlier. Actually, fly is usually enough, but not to be confused with trouser openings where many people hide their genitals.

Genitals?

Oh, yes, many people have genitals.

Do flies have genitals?

Of course, every fly is evidence of a mom and pop fly having genitals.

Then do flies have flies?

Baby flies.

I mean flies to hide their genitals.

Flies are not given to wearing trousers.

Then where do they hide their genitals?

They don't, except maybe on formal occasions such as funerals and weddings when they want to be acknowledged as Musca domestica. But for every day use, housefly will do, it's friendlier. Actually, fly is usually enough, but not to be confused with trouser openings where many flies hide their genitals. . . .

One Lonely Afternoon

Since the fern can't go to the sink for a drink, I graciously submit myself to the task, returning with two glasses of water. And so we sit, the fern and I, sipping water together. . . .

Of course I'm more complex than a fern, full of deep thoughts as I am. But I lay this aside for the easy company of an afternoon friendship.

Yet, had I my druthers, I'd be speeding through the sky for Stockholm, sipping bloody marys with wedges of lime. . . .

And so we sit one lonely afternoon sipping water together. The fern looking out of its fronds, as I look out of mine. . . .

The Farmer Who Lowed

It was possible to grow flesh directly from the earth. Now no need for hooves or horns.

A harvest of boneless blobs lying in the fields like watermelons covered with cowhide; their roots umbilical cords. . . .

Sometimes a throwback like an unfinished memory. One with horns, another with eyes. One lowing secretly inside itself. Sometimes just an udder swollen with milk. . . .

A farmer looking over his fields squints against the setting sun and begins to low. . . .

Of Memory and Distance

It's a scientific fact that anyone entering the distance will grow smaller. Eventually becoming so small he might only be found with a telescope, or, for more intimacy, with a microscope. . . .

But there's a vanishing point, where anyone having penetrated the distance must disappear entirely without hope of his ever returning, leaving only a memory of his ever having been.

But then there is fiction, so that one is never really sure if it was someone who vanished into the end of seeing, or someone made of paper and ink. . . .

Monkey Gas

I ordered ape, and was served monkey. . . .

I had dug right into the corpse curled on my plate like someone dreaming so deeply never to be awakened again, not even by death.

So human, and yet not so much so as to make me out a cannibal. . . .

But, as I was picking ape out of my teeth, and belching what I thought were ape flavored belches, I discovered that I was actually belching monkey gas.

I said to the waiter, Why am I belching monkey instead of ape?

You've probably got a case of monkey gas, he said.

Monkey gas? But the menu said, ape.

It's the octopus, he said, He ran out of ink, and had just enough for ape. Though monkeys are smaller, they take twice the ink.

And so, being true to the above, I continued the belching of monkey gas for no better reason, save that an octopus had run out of ink. . . .

The Rooster's Wife

One day I lay my head in the nest of a hen, hoping to be incubated, and fell asleep. . . .

But as I slept I dreamed an angry rooster cock-a-doodle-dooed at me, What are you doing in my wife's bed?

Please, your highness (I dreamed myself replying), I'm trying to be incubated. Another hen laid me, but I never hatched. Then seeing the possibility of your wife's bosom, without I promise being moved by the slightest erotic possibility inherent in such fullness of breast; nor, might I mention, that cunning little bow of a beak, the sharp, sweet kisses promised therein—I sought only remedy for an unhatchedness. . . .

And then I dreamed I dreamed no more. . . .

ACKNOWLEDGMENTS

Grateful acknowledgment is made to the following journals:

American Letters & Commentary: "The Mare's Egg," "Monkey Gas." *Double Room*: "The Dog's Tail," "The Hollow Pig," "Baby Doll," "To Do No Harm," "Portrait of a Wrinkly Old Man with Nasty Genitals," "Of Memory and Distance." *Fence*: "The Organ of Thought." *Green Mountain Review*: "Geography." *Hayden's Ferry Review*: "The Jack Story," "Thighs." *Jubilat*: "The Farmer Who Lowed." *Margie*: "The Civilized Man," "The Proud Citizen." *NO*: "The Fascination," "The Wound." *Sentence*: "Of Those Who Bring Forth Upon the Earth," "Rocks," "The Knitting," "The Restaurant," "Skeleton." *The Thousands*: "The Rooster's Wife." *Washington Square*: "Of a Midsummer's Night," "Lunch."

ABOUT THE AUTHOR

RUSSELL EDSON is, arguably, the most distinguished American writer of prose poems. His books include, *Appearances* (1961); *A Stone Is Nobody's* (1961); *The Very Thing That Happens* (1964); *The Brain Kitchen* (1965); *What a Man Can See* (1969); *The Childhood of an Equestrian* (1973); *The Clam Theater* (1973); *The Falling Sickness 4 plays* (1975); *The Intuitive Journey & Other Works* (1976); *The Reason Why the Closet-Man is Never Sad* (1977); *With Sincerest Regrets* (1980); *The Wounded Breakfast* (1985); *Tick Tock* (1992); *The Song of Percival Peacock* (1992); *The Tunnel: Selected Poems* (1994); *The Tormented Mirror* (2000); *The House of Sara Loo* (2002); *O Túnel* (2002). Russell Edson lives in Connecticut with his wife, Frances.

BOA EDITIONS, LTD.
AMERICAN POETS CONTINUUM SERIES

COLOPHON

The Rooster's Wife by Russell Edson was set in Century Schoolbook with Bodoni Ornaments by Scott McCarney, Rochester, New York. The cover design was by Steve Smock. The cover art, "Woodcut" by Russell Edson, is courtesy of the author. Manufacturing by McNaughton & Gunn, Ann Arbor, Michigan.

The publication of this book was made possible, in part, by the special support of the following individuals:

John & Lisa Anderson

Jeanne Marie Beaumont

Alan & Nancy Cameros

Wyn Cooper & Shawna Parker

Burch & Louise Craig

Dale T. Davis & Michael Starenko

Suzanne & Peter Durant

Dr. Henry & Beverly French · Dane & Judy Gordon

Howard & Carole Haims · Kip & Deb Hale

Peter & Robin Hursh · Robert & Willy Hursh

Gerard & Suzanne Gouvernet

Archie & Pat Kutz · James Lenfestey

Rosemary & Lew Lloyd

Peter & Phyllis Makuck

Marianne & David Oliveiri

Boo Poulin

Deborah Ronnen · Paul Tortorella

George & Bonnie Wallace

Dan & Nan Westervelt

Pat & Michael Wilder

Printed in the USA
CPSIA information can be obtained
at www.ICGtesting.com
LVHW090806080824
787695LV00003B/389

9 781929 918638